T0216573

DESCRIPTIVE LABELS
FOR BOTANIC GARDENS

BY

HUMPHREY GILBERT-CARTER

Director of the
Cambridge University Botanic Garden

CAMBRIDGE
AT THE UNIVERSITY
PRESS
1924

CAMBRIDGE
UNIVERSITY PRESS

University Printing House, Cambridge CB2 8BS, United Kingdom

Cambridge University Press is part of the University of Cambridge.

It furthers the University's mission by disseminating knowledge in the pursuit of education, learning and research at the highest international levels of excellence.

www.cambridge.org
Information on this title: www.cambridge.org/9781107504820

© Cambridge University Press 1924

First published 1924
Re-issued 2015

A catalogue record for this publication is available from the British Library

ISBN 978-1-107-50482-0 Paperback

In Memoriam

J. B. PEACE

The varnish recommended for labels on waterproof paper on page 7 of this book has proved very uncertain. Though some labels varnished in the manner recommended look fresh after three and a half years exposure, from others recently put out the varnish has chipped away, taking the print with it. The assistants in the Chemical Laboratory have told me a process which seems to give much better results. Photographic film is dissolved in amyl acetate one part and acetone four parts, and the labels immersed in the solution. When they are quite dry they are treated with ordinary paper varnish. Labels on waterproof paper treated with melted paraffin wax have a dull, sodden appearance but a great stability. All labels should be nailed with small copper tacks on to flat pieces of wood.

PREFACE

THE object of this little book is to induce those in charge of collections of plants to exhibit them so as to give information as well as pleasure to as many people as possible.

In the Cambridge University Botanic Garden these labels are printed on green waterproof paper made at the Willesden Paper Mills. In order to protect the printing ink from the weather they are treated abundantly with a syrupy solution of celluloid in either amyl acetate or acetic acid. Experiments are now being made to find out which of these two solvents makes the more lasting varnish.

The idea of making extensive use of descriptive labels occurred to me in March 1921, when I undertook the charge of the wonderful collection of plants at Cambridge, many of which were brought together by that assiduous botanist Mr R. Irwin Lynch, during his forty years Curatorship of the Botanic Garden.

7

PREFACE

The method of printing and protecting the labels has been in use for many years at Kew, and was explained to me by Sir David Prain, at that time Director, shortly after I took up my duties at Cambridge.

The type and arrangement were devised by the late Mr J. B. Peace, University Printer, a faithful friend of the Botanic Garden, who took great interest in the labels. As the Garden could not afford to have them printed he raised a subscription to defray the expenses of preparing three dozen. Including those here published eleven dozen labels have been printed, and because of the kindness of the friends found by Mr Peace, and of others who have joined their ranks, the Botanic Garden has had to pay for none. The expense of printing the labels included here, which have all been prepared since the publication of the Guide in October 1922, and describe plants not mentioned in it, was defrayed by Professor A. A. Bevan of Trinity, Mr F. F. Blackman of St John's, Mr W. Balfour Gourlay of Trinity, The Rev. John Gray of Edinburgh, Dr F. H. H. Guillemard of Gonville

8

and Caius, Mrs John Hopkinson, The Rev. Dr St J. Parry, Vice-Master of Trinity, and Sir Adolphus Ward, Master of Peterhouse.

There are doubtless many other methods of preparing descriptive labels, and all who are interested should try experiments. An inexpensive label, which we have recently found satisfactory at Cambridge, is prepared by typewriting, or writing in Indian ink, on good white card, and afterwards soaking the card in melted paraffin wax. A reproduction of a label of this kind written by Mr H. Tomlinson of Christ's College forms the frontispiece of this little book. Those intending to write labels should study *Writing and Illuminating and Lettering* by Edward Johnston, published by Sir Isaac Pitman and Sons; but handwriting is best suited for labels of only a few words.

Labels describing groups of plants should be in larger letters than those for individual species. The form of such labels depends upon the arrangement of plants in the garden. Our out-of-door collections are arranged *systematically*, and at present have no group labels. Our Temperate House is arranged *geographically,* and will shortly have large group labels describing the various regions represented. As sclerophyllous shrubs abound in the Temperate House a label, of which the frontispiece is a reproduction, describing sclerophyllous vegetation and a map showing its geographical distribution are exhibited on the door.

Outline maps after being filled in can be soaked in melted wax to protect them from water. Enough has been said to show that descriptive labels capable of withstanding water and weather can be prepared in more than one way, and to call attention to the fact that a Botanic or Zoological Garden needs as many labels as a Museum. As a matter of fact botanists need to be reminded that a Latin name affixed to a plant means nothing to the majority of visitors to a Botanic Garden, and that a long row of sclerophyllous shrubs each with its Latin name only produces bewilderment.

A word about ordinary name-labels. The kind we find most satisfactory are made of zinc, and written with a solution of platinum chloride, 15 grains to the ounce. Besides the name of the plant, these labels should always give its family and native land. Of course in collections which are arranged systematically it is not necessary to write the name of the family on each label, as all the plants growing on the same bed belong to the same family, whose name is shown on a large stand-label. It should not be forgotten that facts of human interest excite the curiosity of many visitors more than either the name or family of the plant; for example:—

Lardizabalaceae

LARDIZABALA BITERNATA

Native of Chile, where the stems are used for ropes

This plant grows in our corridor, and near the above label a stem about as thick as a pencil is allowed to ramble in a position calculated to entice visitors to take it in both hands and try to break it. In this athletic seat of learning the plant may suffer harm, but it would be quite safe so labelled and exhibited in any other Botanic Garden.

PREFACE

As most visitors appreciate popular names, an English name should be included on a label whenever a plant has one that is in use. Many English names mentioned in Floras are never used in speech, and some are too misleading to be indiscriminately exhibited in Botanic Gardens. For example the so-called "Berry-bearing Alder" (*Rhamnus frangula*) is not an Alder, neither does it bear berries. But such a name as this gives the label-writer the opportunity of mentioning that the fruit in the family *Rhamnaceae* is *never* a berry, and in the closely allied *Vitaceae* it is *always* a berry.

For the legends of these labels I am responsible, except for several passages, where kind and expert friends, who happily abound in Cambridge, have helped me. They will certainly be found to contain, as all labels must, the results of the writer's idiosyncrasies. They should also contain a consideration for the visitors to this Botanic Garden, who range from grave professors of various subjects both eastern and western, to gay school children. Someone else with other idiosyncrasies and other readers would have told a

12

PREFACE

different tale about the plants, and I would not wish him to say what I have said; but my wish that he should say *something* is the chief reason for making these labels public. Should any of the labels find favour, they can be obtained printed on waterproof paper for a shilling each post free, and the profits of the transaction will go to the Cambridge University Botanic Garden, which always needs all it can get.

Chief among those to whom I am indebted for help are Professor A. C. Seward, Mr A. G. Tansley, and Mr E. J. Thomas of the University Library. Mrs F. F. Blackman furnished the material for the last six lines of the label for *Vitis labrusca* (page 62). Dr Guillemard has very kindly read the proofs.

H. G.-C.

March 1924

Family Cyatheaceae

DICKSONIA ANTARCTICA *Labill.*

Native of Australia and Tasmania. A very closely allied species, *D. fibrosa* Col., is endemic in New Zealand. Like many members of the *Cyatheaceae* it is a "tree fern" with an erect stem, which attains 50 feet in height, surmounted by a crown of graceful fronds. Notice the dense mantle of adventitious roots enwrapping the stem. There is a herbaceous species of *Dicksonia*, *D. apiifolia* Hook., growing in the Corridor nearly opposite the Palm House door. Tree Ferns affect warm and moist localities, and are very abundant in northern South America, the Monsoon Region, East Australia, New Zealand, and the Pacific Islands; in New Zealand they grow close to the Tasman Glacier.

Family Polypodiaceae

ACROSTICHUM AUREUM *L*.

Native of the tropics of both old and new worlds, where it inhabits the shores of tidal estuaries and backwaters, often covering great areas of the brackish mud flats at some distance from the sea. It is the only living fern which can tolerate appreciable quantities of common salt in the soil water. The bright golden yellow of the young pinnae gives the plant its trivial name (Lat. *aureus*, golden). The upper pinnae of the frond are alone fertile; their lower surfaces are completely covered by sporangia. From among the bases of the fronds clusters of roots descend, and enter the mud. This species has many features of anatomical interest, and in several respects resembles other members of the "mangrove vegetation."

16

Family Osmundaceae
TODEA BARBARA *Moore*

Native of South Africa, Australia, and New Zealand. The root-mantled stem tends to grow erect, and in Australian specimens it may weigh as much as a ton and a half. The Cambridge plant is one of the finest specimens in cultivation. The stem, which in the closely allied genus *Leptopteris* is much more slender, gives the plant the aspect of a "tree fern." The great majority of tree ferns belong to the family *Cyatheaceae* (cf. *Dicksonia antarctica*). The only other genus of *Osmundaceae*, besides the two already mentioned, is *Osmunda*, which includes our Royal Fern (*O. regalis* L.). *Osmundaceae* is one of the most ancient families of ferns.

Family Pinaceae
ARAUCARIA EXCELSA *R.Br.*

Native of Norfolk Island (930 miles off the east coast of Australia) : introduced into England in 1793 by Sir Joseph Banks. It is much planted in the warmer parts of the world as an ornamental tree, and its straight stems bearing very regular whorls, usually of five branches, are familiar to visitors to the Riviera, Madeira, Canary Islands, and other places. It is not hardy in Britain, but is very commonly grown as a pot plant. The timber is useful, especially for ship-building. Araucarian foliage-shoots and cones closely allied to those of the Norfolk Island Pine occur in the Jurassic rocks of Yorkshire and in many other parts of Europe. The genus *Araucaria* persisted in Europe to the Tertiary epoch, but is now confined to the southern hemisphere.

Family Pinaceae

PICEA EXCELSA *Link*

COMMON SPRUCE, NORWAY SPRUCE

Native of Central and Northern Europe, but not of Britain. Its remains have been found in English pre-glacial deposits as, for example, the forest bed at Cromer, but not in later deposits. It is very extensively planted on the Continent. Large supplies of the timber ("white deal") come to us from Norway, where the trees grow slowly and make comparatively hard wood. In Germany it is called *Fichte*, and it is the common "Christmas tree" of that country. Burgundy Pitch (*Pix Burgundica*), no longer in the British Pharmacopoeia, is the resinous exudation from the stem. It was formerly much used as a basis for plasters. The cone-like galls usually seen on the Spruce are caused by the insect *Chermes abietis*. The real cones are long and pendulous.

Family Pinaceae
PINUS CEMBRA *L.*

Native of the Alps of Central Europe, where it grows at an altitude of 5–8000 feet, and of Siberia, where it covers large tracts of country at a much lower level. The seeds, known as "pine kernels," are good to eat, and in Siberia thousands of tons of them are collected annually. This species and *P. peuke* Gris., the Macedonian Pine (which has green, glabrous shoots), are the only 5-needled pines in Europe. The Arolla pine is known among other 5-needled pines commonly cultivated in this country by its shaggy young shoots. The closely allied *P. koraiensis* Sieb. et Zucc. has leaves toothed up to the apex. *P. cembra* has an entire leaf-apex.

Family Pinaceae

FITZROYA PATAGONICA *Hk. f.*

Native of West Patagonia and South Chile. This tree was discovered in 1834 during the voyage of the *Beagle*, and was named after Captain Robert Fitz-Roy, the commander of that ship. In 1849 Messrs Veitch's famous collector, William Lobb, introduced it into this country. It is quite hardy in Devonshire and Cornwall, where magnificent specimens are to be seen. Note the white lines of stomata on both upper and lower leaf-surfaces. The only other species of this genus, *F. archeri* Benth., is native of Tasmania, and is not hardy in Britain.

Family Pinaceae
LIBOCEDRUS DECURRENS *Torr.*
INCENSE CEDAR

Native of Oregon and California. It is most abundant and largest on the sierras of Central California at 5-7000 feet above the sea, where it frequently attains a height of 150 feet. In 1853 it was introduced into this country, where it grows with a columnar outline like the Lombardy Poplar. In its native land, however, it forms a very irregular crown. The timber is used for various purposes, but is liable to attack by dry rot. *Libocedrus* is one of the few coniferous genera common to the N. and S. hemispheres. Of the other 7 species 2 grow in Chile, 2 in New Zealand, and 1 each in New Caledonia, New Guinea, and China.

Family Hydrocharitaceae

VALLISNERIA SPIRALIS *L.*

This plant lives rooted to the bottom of ponds in the warmer parts of the world. In deep water its submerged, ribbon-like leaves attain six feet in length. The staminate and carpellary flowers are on separate plants (dioecious). The long, spiral stalks of the carpellary flowers carry them to the surface. The staminate flowers, which arise near the base of the plant, ultimately break loose and drift about on the water till they come into contact with the large stigmas of the carpellary flowers and bring about pollination. The spiral of the carpellary flower stalk then closes and carries the ripening fruit to the bottom of the pond.

Family Gramineae

ARUNDO DONAX *L.*

Native of the Mediterranean Region, where it is often planted to form hedges. This is by far the largest European grass, sometimes attaining twenty feet in height, with stems nearly an inch in diameter. On account of its great stature, and the innumerable uses of its cane-like stems, it has been called the "bamboo of Europe." It is not, however, a bamboo, but is closely related to our common reed, *Phragmites communis* Trin. (*Arundo phragmites* L.). Several words in the O.T. have been thought to signify this plant, which is common in Palestine, and it is probably the "reed (κάλαμος) shaken with the wind" of Matt. xi, 7. The widely creeping root-stocks and roots contain much starch and sugar, and are used medicinally.

BAMBOOS

The tribe *Bambuseae* of the family *Gramineae* (Grasses) contains more than 200 species. They have woody stems, which sometimes reach a great height, and their leaves often have short stalks. The young stems are encased in sheaths, whose blades ("limbs") are small and without midrib. Most bamboos are tropical, but those which are hardy in this country (all of which have tesselated veins in their leaves) come from temperate Asia. Many species die down to the ground on flowering. Nearly all our hardy bamboos with cylindrical stems are placed in the genus *Arundinaria*, and all those in which the internodes are flattened on each side alternately are placed in *Phyllostachys*. Bamboo stems are used for innumerable purposes.

Family Pontederiaceae

EICHORNIA CRASSIPES *Solms*

WATER HYACINTH, MILLION-DOLLAR WEED

Native of Brazil; now naturalised in many hot regions of the world. It usually grows upon water with the lower portion of the leaves swollen to form globular bladders, which cause the plant to float freely. When growing in shallow water or rooted in wet soil the bladders are only feebly developed. The Water Hyacinth multiplies vegetatively with great rapidity, and has become a serious pest in many parts of the world. In India attempts are made to get rid of the plant profitably by burning it for the potash it yields.

Family Liliaceae
BOWIEA VOLUBILIS *Harv.*

Native of South Africa. This plant grows from a large bulb, which produces annually a long, green, slender, twining stem devoid of leaves. In its lower portion the stem gives rise to copiously ramified branches, which represent inflorescences. Note the minute scale-leaves which are present in these parts, and compare them with the bracts of the true inflorescences, which are borne higher on the stem, and produce numerous green flowers. The vegetative form of this plant is probably unique. It might be described as a much branched, fleshy switch-plant.

Family Liliaceae
HYACINTHUS ORIENTALIS *L*.
Common Hyacinth

Native of the Eastern Mediterranean Region. This is an old garden favourite grown by Gerarde in 1596. It has been cultivated for a long period in Holland, and as early as the beginning of the nineteenth century two thousand varieties were distinguished by the Haarlem gardeners. The so-called "Roman" Hyacinths, which bloom earlier, and have fewer, more slender flowers are cultivated in France. They belong to the variety *albulus*. *Hyacinthus romanus* L. is a totally different species. In poetry, especially oriental poetry, curly locks of hair are often likened to Hyacinths, which in Persian are called سنبل :

شكنج گيسوى سنبل ببين بروى سمن – حافظ

Family Liliaceae

PHORMIUM TENAX *Forst.*

Native of New Zealand, Norfolk Island, and the Chatham Islands: introduced into England, where it is fairly hardy, in 1789 by Sir Joseph Banks. The leaves yield an exceedingly tough fibre. In the *Kew Bulletin*, 1919, No. 4 (page 169), is an interesting illustrated account of attempts made to grow New Zealand Flax commercially in the British Isles. The flowers, like those of many other New Zealand plants, are pollinated by birds. Note the handsome, sword-like leaves with bright red keels and margins.

Family Liliaceae

SMILAX ASPERA *L.*

Native of the Mediterranean Region, where it is used in domestic medicine. It is called in French "Salsepareille des pauvres" (true "Sarsaparilla" consists of the dried roots of *S. ornata* Lem., native of Costa Rica). This is the only European representative of the genus *Smilax*, which has about 200 species, most of which are natives of tropical and sub-tropical regions. It is a relic of Tertiary times when tropical plants grew extensively in Europe. Other Tertiary relics in the South of Europe are *Chamaerops humilis*, *Ceratonia siliqua*, *Anagyris foetida*, and *Nerium oleander*; examples of these may be seen in the Mediterranean collection in the Temperate House. Note the net-veined leaves of *Smilax* and the two tendrils which spring from the leaf-stalk.

Family Betulaceae

OSTRYA ITALICA *Spach*

Hop Hornbeam

Native of South Europe, Asia Minor, and Atlantic North America. The American plant (*O. virginica* Willd.) differs from the Old World plant (*O. carpinifolia* Scop.) in its twigs often bearing gland-tipped hairs, in the leaves having fewer lateral veins, and in its larger nut. The foliage bears a confusing resemblance to that of the Hornbeam (*Carpinus betulus*), but the bark, and the carpellary catkins and fruit of the two trees differ widely. Such popular names as the English "Hop Hornbeam" and the French *Charme-houblon* signify a tree like the Hornbeam bearing catkins like those of the Hop. The wood is heavy, close-grained, and very tough.

Family Fagaceae
NOTHOFAGUS OBLIQUA *Blume*
Roble Beech

Native of Chile, where it is a valuable timber tree. The genus *Nothofagus*, of which there are about a dozen species, was formerly included in *Fagus* (true Beeches). They occur in temperate South America and Australasia, and differ widely in appearance from the true Beeches. The leaves are smaller than those of the beeches and are often evergreen. The staminate flowers are solitary, or rarely in threes, never in the globose catkins characteristic of *Fagus*, and there are frequently three nuts in each husk, which is always much smaller than the husks of the true beeches. The southern distribution of the genus is of great interest. No members of the family *Fagaceae* occur in Africa south of the Sahara.

Family Fagaceae
QUERCUS COCCIFERA *L.*
Kermes Oak, Grain Tree

Native of the Mediterranean Region, where in the form of a low shrub it is often dominant on limestone rocks; the arborescent variety *pseudococcifera*, of which Abraham's Oak at Mamre is an example, is abundant in Palestine. The Kermes Oak is the host plant of a scale-insect *Kermes vermilio* Planch. (family *Coccidae*), an ally of the Cochineal insect (*Dactylopius cacti* Costa), which feeds on cactaceous plants. The female insects of both species, when dried, make a beautiful red dye. The Kermes or "Grain" (formerly thought to be seed) of Portugal was celebrated from the time of Pliny (granum...circa Emeritam Lusitaniae in maxima laude est *N.H.* ix, 65) to Chaucer (...for to dyen With brasil, ne with greyn of Portingale). Three sprigs of this tree still form the crest of the Dyers' Company.

Family Ulmaceae

ZELKOVA CRENATA *Spach*

Native of the Caucasus Region. The genera *Zelkova* and *Celtis* belong to the sub-family *Celtidoideae*, which differs from the *Ulmoideae* (elms, etc.) in having stone-fruits. (The winged nuts of the elms are well known.) Compare the bark of the *Celtidoideae* with that of the elms. This tree grows slowly, and lives long; specimens planted at Kew about 1760 are now 60 feet high. In Persia, where it is called درخت آزاد (cf. *Melia azedarach* L., in the Stove), cataplasms of the leaves are used for enlarged glands, and the flowers and a decoction of the leaves are used to destroy vermin.

Family Phytolaccaceae
PHYTOLACCA AMERICANA *L. Spec. pl. ed.* I (1753)
Poke, Scoke, Garget

Native of North America, long grown in Europe, and cultivated by Parkinson and Ray. It is often called by the later name *P. decandra* L. Spec. pl. ed. II (1763). It is poisonous, but its young shoots, rendered harmless by cooking, may be eaten, and are said to resemble asparagus. The roots are poisonous even when cooked. Poke was formerly cultivated in wine-growing districts for its berries, whose crimson juice was used to colour pale wines, and the plant has now become widely naturalised in the warmer parts of the world. Children make red ink from the berries, which, however, have never been used for dyeing. The roots, leaves, and berries are used in domestic medicine, and the root is official in the United States Pharmacopoeia.

35

Family Phytolaccaceae

ERCILLA SPICATA *Moq.*

Native of Chile. This plant climbs by means of adhesive discs, which arise above the leaf-axils. These discs, which when young are thickly beset with processes resembling root-hairs, are not strong enough to hold the plant on a vertical surface, so that if grown on a wall it must be tied up. The genus *Ercilla*, which contains two species (both S. American), is closely allied to *Phytolacca*. Compare the flowers and the almost apocarpous gynaeceum with those of *Phytolacca americana* L., which grows on the bed below this label. In most of the *Phytolaccaceae* the gynaeceum is syncarpous, and in some, e.g. *Petiveria alliacea* L. and *Rivina humilis* L., which grow as weeds in the Stove, there is only one carpel.

Family Cercidiphyllaceae

CERCIDIPHYLLUM JAPONICUM *Sieb. et Zucc.*

Native of Japan and China. It is the largest broad-leaved deciduous tree in both those countries, often growing to 100 feet in height with a trunk of enormous girth. In this country it never exceeds the dimensions of a large shrub. The leaves resemble in form those of the Judas Tree (*Cercis*); hence the name *Cercidiphyllum*. The branches in their second year develop short shoots (spurs), which continue for many years to bear solitary leaves. The Japanese variety of this tree, called *Katsura*, which lives in dense forests, differs widely in habit, though only slightly in structure, from the Chinese variety, which affects glades and park-lands.

Family Lauraceae

CINNAMOMUM CAMPHORA *Nees et Eberm.*

CAMPHOR LAUREL

Native of Formosa, Japan, and China. The name Camphor is applied to several white, odorous, volatile vegetable products having similar properties. This tree is to-day by far the most important source of Camphor, which is obtained by distilling the wood. The Camphor first known to the world was obtained from *Dryobalanops aromatica* Gaertn. (family *Dipterocarpaceae*). In many parts of the world Camphor is extensively used in medicine and perfumery. The Camphor Laurel is hardy in the western counties of England, and is commonly planted by roadsides in South California. The word Camphor is from the Arabic كافور.

Family Lauraceae

LINDERA BENZOIN *Blume*

SPICE BUSH, BENJAMIN BUSH

Native of Atlantic North America. Like most plants belonging to the *Lauraceae*, as for example *Laurus nobilis* and *Umbellularia californica*, the leaves of this shrub are fragrant, and both the above popular names refer to the balsamic aroma they give off when crushed. "Benjamin" is a corruption of Benzoin, the name of a resinous solidified balsam obtained from the incised stems of *Styrax benzoin* Dryand., family *Styracaceae*, which grows in Sumatra. The aromatic bark and fruits of the Spice Bush are used medicinally. This species is called *Benzoin aestivale* Nees by systematists who consider that *Lindera* is the proper name of a genus of *Umbelliferae* known in British floras as *Myrrhis*.

Family Papaveraceae
PAPAVER SOMNIFERUM *L.*
Opium Poppy

Probably a cultivated form of *P. setigerum* DC., which is a native of the Mediterranean Region: extensively cultivated: grown in China from the 8th Century. Opium is the juice obtained by incision from the unripe capsules inspissated by spontaneous evaporation. The Greeks knew opium at the beginning of the third century B.C., and it was probably first prepared in Asia Minor. The Arabs learnt of it from the Greeks, and the Arabic name افيون is from Greek ὄπιον. The spread of Islam distributed opium widely in Asia, and it is now consumed by millions of human beings. Eaten or smoked in moderation it is no more harmful than tea or tobacco, but taken immoderately it is very deleterious. The alkaloid *Morphine* ("Morphia") is obtained from Opium.

Family Rosaceae

COTONEASTER BACILLARIS *Wallich*

Native of the Temperate Himalaya, where it ascends to an altitude of 10,000 feet. The large-leaved species of *Cotoneaster* are often mistaken for "Thorns" (*Crataegus*), from which, however, their entire leaf margins and the absence of spines at once distinguish them. The long, slender branches of this species make excellent walking sticks; hence the trivial name *bacillaris*, from Latin *bacillum*, a small staff or wand. A masculine form of this word (*bacillus*) is used by bacteriologists as a kind of generic name for rod-shaped bacteria. The fruit of this species is almost black when ripe. The allied *C. frigida* Wallich (also Himalayan) has bright red fruits.

Family Rosaceae
PYRACANTHA COCCINEA *Roemer*
PYRACANTH

Native of South Europe, Asia Minor, and the Caucasus; introduced in 1629 into this country, where it is often planted on account of its graceful foliage, copious white flowers, and beautiful orange-red fruits. This and several allied species, all of which have evergreen, obscurely serrate or almost entire leaves, and leafy thorns, are now generally considered to constitute the genus *Pyracantha*, a group of plants which is, however, often included in other genera. *Crataegus*, in which it is sometimes placed, is distinguished by having deciduous, lobed leaves, and leafless thorns, and *Cotoneaster* by its quite entire leaves and absence of thorns.

42

Family Rosaceae
CYDONIA VULGARIS *Persoon*
COMMON QUINCE

Probably native in South Europe; long cultivated in many countries for its yellow, fragrant fruit, which is unfit to eat raw, but makes good preserves. In Portuguese it is called *marmelo*, whence our "marmalade," which was originally made of quinces. Various parts of this plant are used medicinally. The seeds contain $20°/_{o}$ of a gum (bassorin), which they readily impart to cold water forming a thick jelly used as a mucilage for toilet preparations. An infusion of Quince seeds is official in several continental pharmacopoeias. The Quince is used as a stock for grafting Pears.

Family Rosaceae

CRATAEGUS MONOGYNA *Jacq.*

HAWTHORN, WHITETHORN, MAY

Native of Europe, and from the Mediterranean Region to the Himalaya. Of our native shrubs hawthorn is one of the commonest and most tolerant, growing alike on chalk, sand, clay, and on the drier parts of fens. It is often the only shrub found on pastured grass land, doubtless because its spines protect it from grazing animals. In Great Britain it is by far the commonest hedge-plant. "Quick" hedges are so closely pruned that the Hawthorn attains no size, and seldom flowers, but it is naturally a graceful, round-headed tree 20 or more feet high. This species differs from our other *Crataegus* (*C. oxyacantha* L.) in its taller stature, more deeply cut leaves, hairy flower-stalks, single style, and smaller, rounder fruits.

Family Rosaceae
CERCOCARPUS PARVIFOLIUS *Nutt.*
Mountain Mahogany

Native of the mountain ranges of Pacific North America, and very common in California. The genus *Cercocarpus* is allied to *Dryas* and *Geum*, but the flowers have no petals, and there is only a single carpel. The fruit is an achene included in the calyx-tube, and tipped (like that of *Dryas*) with the elongated, persistent style, which is thickly beset with long, white hairs. The word *Cercocarpus* (from κέρκος, tail, and καρπός, fruit) refers to this curious, long-tailed fruit. The hard, brittle wood of all the species makes good fuel, and is sometimes used in the manufacture of small articles.

45

Family Leguminosae

MIMOSA PUDICA *L.*

THE SENSITIVE PLANT

The Sensitive Plant is said to be a native of Brazil, but it now grows throughout the tropics of both hemispheres. The leaf has two or four pinnate leaflets, which radiate from the end of a long stalk or petiole. Contact, concussion, or darkness causes collapse of the leaves, which takes place in three stages. Firstly the tiny pinnules move upwards and slightly forwards till they close together in partially overlapping pairs. Secondly the petiolules close together, and thirdly the whole leaf droops downwards. Several species of *Mimosa*, and many other plants, show similar leaf movements. In India this plant is used medicinally. Its Hindustani names لجوَتی and لجالو mean "bashful," and چھوئ موئ means "if you touch me I die."

Family Leguminosae
CERATONIA SILIQUA *L.*
Locust Tree, Carob Tree

Native of Arabia; long cultivated, and perhaps wild, in other parts of the Mediterranean Region. The hard, lustrous, reddish wood is used for marquetry, and the tree is grown in Algeria for walking sticks. The sugary unripe pods are the Locust Beans used for fattening cattle, and eaten by the poor. They were the husks ($\kappa\epsilon\rho\acute{a}\tau\iota a$) of the parable of the prodigal son (Luke xv, 16). The seeds were the original "carat" weights used by jewellers. The word carat is from Arabic قيراط, perhaps from Greek $\kappa\epsilon\rho\acute{a}\tau\iota o\nu$, meaning the pods, diminutive of $\kappa\acute{\epsilon}\rho a\varsigma$, a horn. The French *Caroubier*, and other similar names, are all ultimately from the Arabic خروب. The Spanish *Algarrobo* applied to this tree, and to *Prosopis alba* in South America, is the same word with the Arabic article prefixed.

Family Leguminosae

PIPTANTHUS NEPALENSIS *D. Don*

Native of the Temperate Himalaya from Simla to Bhotan, at an altitude of 7-9000 feet: introduced into this country in 1821. The stipules are united to form a sheath. Note that the stamens are not united as they are in most *Papilionatae*. In two tribes of *Papilionatae* the stamens are free. In the *Podalyrieae*, to which this shrub belongs, the leaves are simple or palmate. Species of allied herbaceous genera, *Baptisia* and *Thermopsis*, grow on the Herbaceous Beds. In the other tribe with free stamens, the *Sophoreae*, the leaves are pinnate. To this latter tribe belong *Sophora* and *Cladrastis* (on the Cornel Plot), and *Myroxylon* (in the Stove). Examine the stamens of *Laburnum* (tribe *Genisteae*), with which *Piptanthus* is often wrongly associated.

Family Leguminosae

LABURNUM ANAGYROIDES *Medik.*

COMMON LABURNUM, GOLDEN CHAIN

Native of Central and Southern Europe, affecting woods on calcareous mountains. This species and *L. alpinum* Presl (see below) have been cultivated for centuries in Britain on account of their pendulous racemes of bright yellow flowers. They produce abundant seedlings and would probably become naturalised were it not for the avidity with which rabbits destroy young plants. The heart-wood of the Laburnums is remarkably hard and dark, and takes a fine polish. The seeds, bark, and wood are poisonous to human beings.

L. alpinum Presl, the so-called Scotch Laburnum, also native of Central and Southern Europe, has larger leaflets with nearly glabrous surfaces and ciliate margins, longer racemes produced two or three weeks later, and glabrous pods with thin, sharp upper margins.

Family Leguminosae

SPARTIUM JUNCEUM *L.*

Native of the Mediterranean Region, where it prefers calcareous soils. The leaves of this shrub soon fall, their function then being performed by the green, rush-like stems. Such plants are called switch-plants, and are abundant in the Mediterranean Region, especially among the *Leguminosae* of that district; but switch-plants occur in many different families in diverse parts of the world, generally in dry habitats: cf. *Ephedra* (beside Middle Walk), *Casuarina* (in Temperate House), and *Polygonum equisetiforme* (in Bay No. 6). Our Broom, *Sarothamnus scoparius* Koch, found especially on sandy soil, is a switch-plant. The stems of the Spanish Broom yield forage and fibre, and the flowers a yellow dye. Sprays of its large, fragrant blossoms are strewn on the ground for religious processions.

Family Leguminosae

CARMICHAELIA FLAGELLIFORMIS *Col.*

Native of New Zealand: fairly hardy in England. The genus *Carmichaelia* consists of twenty species, which are all confined to New Zealand except one (*C. exsul* F. Muell.), which grows only in Lord Howe Island. Many of them, like this specimen, are "Switch-plants," i.e. their branches are long, slender, and green, and perform the work of the leaves, which are minute, and fall early. The species of *Carmichaelia* are difficult to discriminate, and sometimes the branches produced by the same plant are flattened in Spring and nearly cylindrical in Autumn. The pod has a persistent "replum" like that of the *Cruciferae*.

Family Leguminosae

PSORALEA BITUMINOSA *L.*

Pitch Clover

An abundant weed in the maritime zones of the Mediterranean Region, Madeira, and Tenerife. All who have visited those parts remember how the air is often filled with its penetrating, bituminous smell. It is used in domestic medicine, and in former Pharmacopoeias was called Herba Trifolii Bituminosi. In Tenerife it is one of the few herbs that remain green throughout the long dry summer, so that it is a very valuable fodder for goats. The genus *Psoralea* is not allied to the true clovers (*Trifolium*), but belongs to the tribe *Galegeae*. Many species of *Psoralea* have pinnate leaves.

Family Leguminosae

ARACHIS HYPOGAEA *L.*

Pea Nut, Monkey Nut

Native of Brazil: now cultivated in nearly all tropical countries, and occasionally in the South of Europe, for the sake of its pea-like seeds ("monkey nuts"), which contain an oil used chiefly for soap-making and lubrication. Large quantities of this oil are used to adulterate Olive oil. After fertilisation the flower-stalks bend downwards, and lengthen, forcing the ripening pod into the soil. Examples of plants in this country that bury their fruits are: *Linaria cymbalaria* (the Ivy-leaved Toad-flax), and *Trifolium subterraneum* (the Subterranean Clover).

Family Leguminosae
ERYTHRINA CRISTA-GALLI *L.*
CORAL TREE

Native of Brazil. The genus *Erythrina* has about fifty species, nearly all of which are tropical. It belongs to the tropical tribe *Phaseoleae*, which are predominantly twiners, the leaflets of whose trifoliate leaves are furnished with *stipels*, which correspond to the *stipules* of the leaves (a familiar example is *Phaseolus multiflorus* Willd., the Scarlet Runner). The flowers of *Erythrina* are resupinate (upside down), so that the large conspicuous standard is beneath the keel, which encloses the stamens and carpel. In the flowers of this species the "wings" are rudimentary.

Family Zygophyllaceae
PEGANUM HARMALA *L.*

Native of the Mediterranean Region and Western Asia, and a very well-known medicinal plant in the East, where it is called سِپَند, اسپند, and حَرمَل, words which are often translated "Wild Rue." (The true Rue, *Ruta graveolens* L., family *Rutaceae*, in Islamic medical works is called سُداب.) All parts of the plant are used medicinally, especially the seeds, which contain two alkaloids and a resin. They yield also a dye, which was at one time tried in Europe, but it could not compete with the aniline dyes. In Persia the seeds are sprinkled upon burning coal in order to avert the evil eye. This and other popular uses are often referred to in Persian books.

Family Rutaceae (*tribe* Diosmeae)

COLEONEMA ALBUM *Bartl. et Wendl.*

Native of south-west Africa, and common on the hills about Cape Town. A large proportion of the plants composing the south-west African sclerophyllous vegetation, as for example many of the members of the above tribe, have very narrow leaves, which are either acicular (needle-like) or ericoid (flat with margins rolled backwards). The genus *Coleonema* has four species characterised by the presence of five staminodes, each enclosed in a sheath formed by the claw of one of the petals. (κολεός, sheath, and νῆμα, thread.) Like the *Myrtaceae* the *Rutaceae* are aromatic from the presence of volatile oil contained in glands, which appear as pellucid dots on the leaves. (Cf. *Citrus*, *Ruta*, and *Skimmia*.)

Family Meliaceae

CEDRELA SINENSIS *Jussieu*

CHINESE CEDAR, CHINESE MAHOGANY

This is a very common tree in central and western China, where the young shoots are eaten as a vegetable, and the timber, which is beautifully marked, durable, and takes a good polish, is used for innumerable purposes. Its bark, separating in long strips, differs widely from that of the Tree of Heaven, with which it is often confused. All the British representatives of the vast order *Geraniales* are herbs; but most of the members of the exotic families *Simarubaceae* (e.g. *Ailanthus glandulosa*, the Tree of Heaven), *Burseraceae*, and *Meliaceae*, and many belonging to the *Rutaceae* and *Euphorbiaceae*, are trees with pinnate leaves, and often resemble the Chinese Cedar in bearing their foliage in tufts at the ends of the branches.

Family Meliaceae

MELIA AZEDARACH *L.*

PERSIAN LILAC, WEST INDIAN BEAD TREE

Native of Central and Western China, where it is very
common : now widely distributed and naturalised in the
warmer parts of the world. The stones of the fruits are often
used as beads for necklaces and rosaries, whence such names
as Bead Tree and *Arbor sancta*. Notice the bi-pinnate leaves
characteristic of the genus *Melia*. In all other *Meliaceae*
the leaves are either simply pinnate, or (more rarely) simple.
Notice also the staminal tube found in nearly all *Meliaceae*. In India, where it is very common, various parts are
used medicinally. Common Indian names are بَكايِن and
كِهوڑا نيم. The word *azedarach* is from the Persian آزاد درخت,
meaning free, or noble tree, applied to this and other trees.

Family Euphorbiaceae

EUPHORBIA CANARIENSIS *L.*

Native of the Canary Islands, where it grows in huge masses on rocks and cliffs from sea level to 300 m. The spurges (Euphorbias) of the section *Euphorbium* differ from our British species in being shrubs or trees with stout, succulent, often cactus-like stems. This species belongs to the sub-section of *Euphorbium* called *Diacanthium*, which is characterised by bearing thorns in pairs, or more rarely in threes, on the ridges of the stems. The *Diacanthium* Euphorbias superficially resemble *Cereus* (family *Cactaceae*), from which they may be distinguished by their milky juice (which is dangerously acrid) and stout thorns arranged in pairs or in threes. *Cereus* has watery juice and *tufts* of thorns mixed with hairs. See the *Cereus* collection at the farther end of Second Succulent House.

Family Anacardiaceae

SCHINUS MOLLE *L*.

PEPPER TREE

Native of Pacific Tropical S. America. The Pepper Tree is extensively planted in warm countries, especially in California and in the Mediterranean Region. Its large, round crown, consisting of graceful, pendulous branches, gives it a superficial resemblance to the Weeping Willow. In California it is losing popularity because it harbours the black scale so fatal to citrous fruits. All parts of this tree abound in resinous substances, which give the crushed leaves a refreshing aroma. The ripe fruits are about the size of peppercorns, which they resemble in flavour, but the tree is not related to the true Peppers. The resin which exudes from the stem is called American Mastic.

Family Vitaceae
VITIS VINIFERA *L.*
Grape Vine

The Grape Vine is probably a cultivated form of *Vitis sylvestris* C. C. Gmel., which is wild in the region of the Danube, from the Mediterranean Region to Central Asia, and in parts of France and Germany. From remote antiquity men have prepared wine from grapes, and in Homer the phrase σῖτος καὶ οἶνος (or μέθυ), meaning bread and wine, occurs frequently. Many ancient place-names, as Ἄμπελος and Οἰνοῦσσαι, refer to vineyards and wine. Vineyards in this country are mentioned in the earliest Saxon charters, but good vintage needs the dry, warm summers of the Continent. The tendrils of the Grape Vine are opposite to the leaves; but every third leaf is without a tendril (cf. *V. labrusca*).

Family Vitaceae

VITIS LABRUSCA *L.*

NORTHERN FOX GRAPE

Native of Atlantic North America. This species is readily distinguished from the Vine (*Vitis vinifera* L.) by its obscurely lobed leaves, each of which stands opposite either to a tendril or to an inflorescence. Many varieties are cultivated for their fruit in America. As its roots are immune from the attacks of the dreaded *Phylloxera vastatrix* it is much used in Europe as a stock for grafting the Grape Vine. Leif the Lucky (who landed in America in the year 1000 A.D.) and subsequent Vikings, called the country "Vínland" because of the presence of this or a similar plant. Tyrker the German found "wine-wood and wine-berries" there, and knew them because he had grown up in a vine-country.

Family Myrtaceae

LEPTOSPERMUM SCOPARIUM *Forst.*

TEA TREE, MANUKA

Native of Australia, Tasmania, and New Zealand : hardy in the milder parts of the British Isles. It is the commonest shrub in New Zealand, where it grows gregariously, and flowers so freely at Christmas that the landscape appears covered with snow. Like all other *Myrtaceae* it is aromatic from the presence of volatile oil contained in glands, which appear as pellucid dots on the leaves (cf. *Eucalyptus, Myrtus, Pimenta,* etc.). The early colonists used the leaves for making tea. The wood is useful for fences and firewood, and the Maoris used it for paddles and spears. The twigs make excellent brooms.

Family Myrtaceae
EUCALYPTUS GLOBULUS *Labill.*
Blue Gum

Native of Tasmania, Victoria, and New South Wales. This tree has been extensively planted in warm countries, and is supposed to improve malarial districts. The leaves of young plants, and those on adventitious shoots of adult trees, are opposite, sessile, and ovate in outline. The adult leaves are alternate, stalked, much narrower, and often sickle-shaped. The timber is useful for many purposes, and, when stained, resembles mahogany. The well-known "Oil of Eucalyptus" is distilled from the fresh leaves of this and other species of *Eucalyptus*, and Red Gum or "Eucalyptus Kino" is the astringent exudation from the stem of various species.

Family Ericaceae

ERICA ARBOREA *L.*

Native of the Mediterranean Region, Madeira, Tenerife, the Caucasus, and the lofty mountains of tropical Africa. In Tenerife it attains the height of fifty feet. In the Western Mediterranean Region it is often associated with *E. scoparia* L., which has glabrous twigs and greenish flowers; and the dried branches of both these species are used for making brooms, and screens against the wind and sun. "Briar" or "brier" tobacco pipes, which appeared in this country in 1859, are made of the roots of this tree. The word, used in this sense, is a corruption of the French *bruyère* applied to *Calluna vulgaris* L. (*bruyère commune*), and to all the French species of *Erica*, *E. arborea* being called *bruyère arborescente*.

Family Oleaceae

JASMINUM OFFICINALE *L.*

Common Jasmine

Native of Kashmir, Afghanistan, Persia, north-west India, and China. This climbing shrub, which grows with remarkable rapidity, has long been cultivated in English gardens on account of its graceful, pinnate leaves and deliciously fragrant white flowers. It is naturalised in the South of Europe, where it grows in hedges, thickets, and rocky places. In the East it is called یاسمین, and many variants of this word occur frequently in Persian poetry:

گفتمش پیراهن از برگ سمن نازکتر است

گفت اندامم نگر کز پیرهن نازکتر است — کمال اصفهانی

Family Sapotaceae

ARGANIA SIDEROXYLON *Roem. et Schult.*

Native of south-west Morocco, where it forms vast forests. In March the Moors collect large quantities of the fruits, from which they strip off the husk, and feed it to camels, goats, sheep, and cows. These animals eat it with avidity; but horses, asses, and mules will not touch it. The stones of the fruits are broken to extract the seeds, from which is expressed "Argan Oil," which the Moors take with their food, as the south-Europeans do Olive oil. One thousand hundredweights of this oil were formerly consumed annually in the Argan district, and probably as great a quantity is still used there. The wood, which is heavy, tough, and fine-grained, is used for various purposes. *Argania* is from the local name ارجان.

Family Labiatae
ROSMARINUS OFFICINALIS *L.*
ROSEMARY

Native of the Western Mediterranean Region and Asia Minor; especially abundant on calcareous soils. The leaf has the texture of a sclerophyll, but approaches the ericoid type in its narrow outline, revolute margins, and felted lower surface. The word Rosemary is from Latin *ros marinus*, "sea-dew," assimilated with Rose and Mary. The Arabs aptly name it اكليل الجبل ("crown of the mountain"), for it affects heights, and in company with *Juniperus oxycedrus* L., *J. phoenicea* L., and *Quercus ballota* Desf., forms the last outposts of Mediterranean vegetation near the summits of the Saharan Atlas at 6500 feet. Rosemary was formerly much used medicinally, and its volatile oil is an ingredient of Eau de Cologne, and of many hair washes.

Family Labiatae

LAVANDULA SPICA *L.* (*L. vera* DC.)

COMMON LAVENDER

Native of the Mediterranean Region, where it affects the montane zone. Together with Box (*Buxus sempervirens* L.) and *Genista cinerea* DC. it covers many square miles of the disforested limestone hills of Provence at 800—1500 m. on their southern exposures. This plant community is called "montane garigue." Lavender has been grown in our cottage gardens for centuries. Its deliciously fragrant flowers are used to perfume clothes and to keep away moth. From the flowers of this species is distilled "Oil of Lavender" of the British Pharmacopoeia. Note the broad, 7-nerved, acuminate bracts. *L. latifolia* Vill. (sometimes, though wrongly, called *L. spica*) has broader leaves and narrow, nerveless bracts. From it is distilled "Oil of Spike" which is used in the preparation of certain varnishes.

Family Labiatae

LAVANDULA STOECHAS *L.*

Native of the Mediterranean Region, Madeira, and Tenerife: a well-known medicinal plant from ancient times. Dioscorides called it στοιχάς, and said that it grew on islands having the same name (Στοιχάδες was the old name for Les Iles d'Hyères). The dried plant is common in eastern drug shops under the name of اسطوقودس, اسطُوخُودُوس, etc., adaptations of the Greek in an oblique case with the Arabic article prefixed. It was used in India for Influenza during the great epidemic of 1918. In the مخزن الادويه (*Treasury of Drugs*) it is described as "The broom of the brain, sweeping away all peccant, phlegmatic, effete humours, removing obstructions, strengthening, dissolving corrupt crudities, and rarefying the intellect." The upper bracts form a purple tuft on the top of the inflorescence.

Family Labiatae
HYSSOPUS OFFICINALIS *L.*
HYSSOP

Native of South Europe and West Asia. It has been long cultivated in most parts of Europe, and was formerly much used in medicine. Until quite recently hyssop tea was given for colds. The word Hyssop is an adaptation of the Hebrew אֵזוֹב, but there is diversity of opinion about the plant intended by this word, and some consider it to signify the Common Caper (*Capparis spinosa* L.). Theophrastus uses the word ὕσσωπος for a plant whose identity is unknown, and in the Septuagint the Hebrew אֵזוֹב is always translated by this word, which in the Vulgate was rendered *Hyssopus*: "Asperges me *hyssopo*, et mundabor," Ps. LI (L).

71

Family Caprifoliaceae
VIBURNUM LANTANA *L.*
Wayfaring Tree, Mealy Guelder Rose

Native of Europe, North and West Asia, and North Africa. In Britain it is native from Yorkshire southwards, and is especially abundant by roadsides, whence its name, and at the edges of woods, on calcareous soil. It is a characteristic and common member of chalk and limestone scrub and coppice. It may be known among our shrubs by its opposite, broad, wrinkled, finely serrate leaves, and scurfy pubescence. This "scurf" on the young stems and leaves is seen under a lens to be made up of numerous star-like ("stellate") hairs. The hard, white wood is good for turnery, but it has an unpleasant smell.

Family Caprifoliaceae

VIBURNUM OPULUS *L.*

GUELDER ROSE

Native of Europe (including Britain), North and West Asia, and North America. The opposite, palmately-lobed leaves resemble those of a Maple; but note the stipules, which are absent in the Maples. The outer flowers of each inflorescence are sterile, and have large, conspicuous corollas. The "Snow-ball Tree," often planted in gardens, is the var. *sterilis* DC., in which all the flowers are of the sterile kind, and are arranged in globular inflorescences. This variety possesses great beauty of flower, but has the disadvantage of never ripening the bright red, translucent fruits, which adorn our country side in autumn. These fruits contain valerianic acid, but are eaten in Norway and Sweden with honey and flour.

C. 73 10

Family Caprifoliaceae
VIBURNUM TINUS *L.*
LAURUSTINUS

Native of the Mediterranean Region. In the South of Europe it abounds locally, especially in rather sheltered valleys and ravines near the sea. This habitat is in accordance with the leaf form, which is intermediate between the sclerophyll and the laurel types. Laurustinus is often grown in English gardens on account of its handsome evergreen foliage and masses of beautiful fragrant flowers, which bloom between the Autumn and Spring. It will stand considerable frost, and shoots freely after severe cutting back. The drupes, which are at first blue, and blacken on ripening, are used as a remedy for dropsy; but their action is very drastic.

Family Cucurbitaceae

LAGENARIA VULGARIS *Ser.*

BOTTLE GOURD

According to De Candolle this gourd, which is cultivated throughout the hotter parts of the world, is native of India, the Moluccas, and China. The young fruit is eaten boiled like vegetable marrow, put into curries, or sliced and boiled like French beans. The dried shell of the mature fruit is used for holding fluids, and is made into musical instruments. It can be made to assume various forms by tying string round it while growing. The leaves, fruit, and seeds are used medicinally. Common Indian names are : كدُو and تُمْڑی) نُوکی is a name for the wild kind). The word *Lagenaria* is from Latin *lagena*, a flagon. There are many cultivated varieties with curiously shaped fruits.

Family Compositae

SENECIO CINERARIA *DC.*

Native of the Mediterranean Region, where it often grows in abundance near the sea, whence Linnaeus called it *Cineraria maritima*. It is naturalised in England and Ireland, and is spreading rapidly in some seaside localities. Hybrids between this species and *S. jacobaea* are described in the *Journal of Botany*, XL (1902), p. 401. The word Cineraria is from Latin *cinerarius*, pertaining to ashes, and the name refers to the ash-coloured down on the leaves of this and other species. It is one of the many plants which in gardens are called "Dusty Millers." The genus *Cineraria* is now usually included in *Senecio*. The garden "Cinerarias" are derived from *Senecio cruentus*, a native of the Canary Islands, and other Canarian species.

INDEX

INDEX

78

INDEX

79

INDEX

Printed in the United States
By Bookmasters